Order this book online at www.trafford.com
or email orders@trafford.com

Most Trafford titles are also available at major online book retailers.

Print information available on the last page.

ISBN: 978-1-4907-6018-6 (sc)
        978-1-4907-6017-9 (e)

Library of Congress Control Number: 2015907343

Because of the dynamic nature of the Internet, any web addresses or links contained in this book may have changed since publication and may no longer be valid. The views expressed in this work are solely those of the author and do not necessarily reflect the views of the publisher, and the publisher hereby disclaims any responsibility for them.

Our mission is to efficiently provide the world's finest, most comprehensive book publishing service, enabling every author to experience success. To find out how to publish your book, your way, and have it available worldwide, visit us online at www.trafford.com

Any people depicted in stock imagery provided by Thinkstock are models,
and such images are being used for illustrative purposes only.
Certain stock imagery © Thinkstock.

Trafford rev. 05/15/2015

www.trafford.com

North America & international
toll-free: 1 888 232 4444 (USA & Canada)
fax: 812 355 4082

CHECK THIS OUT

A Brief Look at World Mysteries and Rarely Told Histories

AMON GOODEN, B.S.

To all those who seek the truth.

Contents

Foreword vii

Preface x

Acknowledgments xi

Introduction xiii

1   The Universe 1

2   Venus 3

3   Mars 5

4   Europa 7

5   The Klerksdorp Spheres 8

6   Piri Reis Map 9

7   Machu Picchu 10

8   Tiahuanaco (Tiwanaku) and Puma Punku 12

9   Nazca Lines 13

10 Jupiter Temple and the Stone of the South 14

11 Moai 15

12 Pyramids at Giza, Egypt, Africa 17

13 Saqqara Bird 19

14 Ancient Model Planes 21

15 The Olmec Culture 22

16 Tombstone of Pical 24

17 Utah Petroglyphs 25

18 The Great God Mars 26

19 Uzbekistan Cave Painting 26

20 Dropa Discs 27

21 Sumerian Culture 28

22 Closing Remarks 30

Sources 32

I can assure you that flying saucers, given that they exist, are not constructed by any power on Earth.

—President Harry S. Truman, 1950,
http://www.ufoevidence.org/documents/doc1358.htm

Foreword

Amon Gooden was born in Cincinnati, Ohio, and has lived in Lorain, Ohio; Louisville, Kentucky; West Chester, Pennsylvania; and Atlanta, Georgia. Amon Gooden graduated from Lincoln University in Pennsylvania with a bachelor's degree in physics with minors in mathematics and preengineering. While a student at Lincoln, Amon worked as a mathematics and physics tutor. Amon was also on the Honda Campus All-Star Challenge National Team and was selected from thousands of students to represent the Honda Motor Company in its national ads. Amon appeared in several issues of *Ebony*, *Black Enterprise*, and the *Crisis Magazine*.

Amon has been working on his book, *Check This Out*, for several years. *Check This Out* is especially critical for Black children and students who hate school because they do not feel a connection to the information. *Check This Out* is also important to Black adults who have grown up believing that being Blacks and of African ancestry is bad. Many Blacks want to be anything but Black and work diligently to hide any characteristics that are a reminder of African heritage. Blacks, from all over the world, have been educated by people who have been teaching that people of European ancestry are the primary people who have created advanced civilizations. Asians are given some credit for impressive contributions, but Blacks are totally omitted from the annals of history as major contributors to world civilizations. If you go to any museum, you will find African history separated from Egyptian history—as if the founders of Egypt (Kemet/KMT) were not Black. Blacks were the founders of the most amazing civilizations on Earth. We are proud that our son has been inspired to research and present to you the truth.

Warren E. Gooden, PhD
Ayo Maria Casey Gooden, PhD, ABPBC

To my mathematical brain, the numbers alone make thinking about aliens perfectly rational.

—Stephen Hawking

Preface

First, thank you for taking the time to read my book. *Check this Out* is the first book in a line of books I have decided to write. My decision to write *Check this Out* is the result of all the things I have experienced and learned. Because of my degree in physics, many of the people whom I know have asked me about my opinions on the history that most people have been taught about the possibility of life on other planets. *Check This Out* is the result of several years of research analyzing different sources of information and examining the origin of people on Earth and major scientific accomplishments.

Acknowledgments

First of all, I would like to thank the universe for being and all the ancestors who paved the way. I am forever grateful for my parents, Dr. Ayo and Dr. Warren Gooden. They are the reason I am an educated and curious person. Special thanks to my mother for giving me the idea to create a book in the first place, and to my father for giving me some of the most important scientific information. It is truly a privilege to be their child. I also want to thank my grandmother, Marilee Casey, and my grandfather, Vibert Gooden, and my uncles, Lawrence Gooden and Morris Downs, Jr., for their wisdom and guidance.

Thank you to all who have taught me so much throughout my life. My brother, Brandon Gooden, introduced me to a lot of new material. Halston Jackson has been one of my best friends for most of my life and has been a gateway to much of the information I have acquired. I want to thank my girlfriend, Tiffany Moore, for supporting and being patient with me. I also want to name the friends who have always been there for me and have had the largest impact on me. Membere Abraham, Walter Anderson, Casandra Basciano, Lawrence Box, Jennifer Brady, Daniel Cherrin, Ashley Coleman, Grachaun and Carlos Correa, Jazmyn Couram, Barry Cropp, Steven DeShields, Jason Douglas-Bowers, Johnny Duck, Femi, Louri and Brandon Fitzgerald, Ed Fitzgerald, Corey Floyd, Foluso, Dr. Donald Ford, Katrina Francis, Jameyshia and Jimmy Franklin, Kent Guerrier, Andrew Haman, Eric Horton, Ino, Mylea Jackson, Dwight Johnson, Tryphenia Johnson, Courtney Justice, Candace and Walter Kellogg, Daniel Lee, Durrell Lee, Masiira Abdul-Malik-El, Jasmine Moore, Kenyatta Moore, Tammy and Jerry Moore, Tudie Nanco, Amber Newton, Gabby Newton, Courtney Patterson, Alisa Pearson, Yaron Pickens, Quron, Ali Shabazz, Jennifer Socarras, Curry Stewart, Michael Stokes Jr., Todd, James Townsend, Tiffany Vann, Anthony Washington, and Lamel Wright.

Family is important and should never be neglected. The family members I want to recognize are Ella and John Hunt, Carmelita Casey-Rubio, Lorenzo Casey, Nicardo Casey, Florence Johnson, Stephanie Fitzpatrick, Conroy Gooden, Diane House, Brandon House, Lauren and Rob Jones, Shaunda Kelly, Kimberly Kelly, Angelique Kenzer, Ariana Kenzer, Derrick Kenzer, Kasey Kenzer, Micah Kenzer-Crawford, Sally Kenzer, Tomorrow Kenzer-Chapel, Amber and Floyd Dixon, Courtney King, Monica King, Barbara Mullett, Donna

Mullett, Dr. Donald Mullett, Jamal Owens, Kelli Owens, Lynn Owens, Shelli Owens, Carmelita and Tony Pierce, Tony Pierce Jr., Sharon Kelly, Herronald Kelly, Sam Kelly, Rose Gooden, and Mulea Wambua.

When I was in school, there were teachers who I felt truly cared and helped me succeed. I would like to thank my math teacher Dr. Abdulalim Shabazz, PhD, my physics teacher Dr. Lynn Roberts, and my chemistry teacher Dr. Robert Langley. My kung fu teacher, sifu Billel (Bill Witacker), has provided me with so much knowledge I will be eternally appreciative.

Unless we are born wealthy, we end up having to get jobs. I want to recognize those people who assisted me in acquiring employment. Dr. Reginald and Mrs. Yolanda Banks, Ms. Rachel Manson, Dr. Malcolm Riley, III, Ms. Lametha Northern, Ms. Gladys Ramirez-Wrease, Mr. Ashmel and Mrs. Maria Williams, and Mrs. Virgie Sangyang.

I apologize for not being able to list everyone I know, but these are the ones who had the largest impact on me so far.

Introduction

I know that, at this very moment, there are radio signals from other civilizations passing through this room which we could detect and study if we but pointed our antennas in the right direction and tuned to the right frequency.

—*Frank Drake*

From the day we are born, we are taught how and what to think by those around us. Most people are unaware of the misinformation that they have been receiving throughout their lives. The mainstream public gets the majority of their information from either the media or the education system. These two sources are biased and have changed and omitted much of the true history and information available. The truth is not taught to the masses because those who are in control do not want to give up their control. Truth is the light, and knowledge is power. By keeping true knowledge a secret, the common person is kept powerless. Those in power have not just hidden information from us, they have convinced the majority, of people in the world, that much of the ancient knowledge and science is just a story—a fairy tale. In order to discredit the truth even further, movies, shows, and books use actual or possible events and knowledge, but call the works "fictional."

The history and knowledge of people of African lineage have been omitted and/or changed more than any other ethnicity. All the wondrous ancient megalithic structures still visible today were created by Black Africans or beings. At the time when sites such as the Great Pyramids at Giza, Tiahuanaco, the Jupiter Temple, and many others were built, the Caucasian had only recently evolved from their African ancestors. Scientists estimate that the first Caucasians emerged from the Caucus Mountains in Asia between twenty and forty thousand years ago. The first civilizations on every continent were therefore created by Black-skinned people. So-called experts are beginning to admit that the age of ancient sites are far older than they were previously thought to be. It is possible that these Black people not only explored this planet but also traveled to other planets as well. The Black people that built the earliest civilizations on Earth might have actually originated on a distant planet and might have brought life here to Earth.

Check This Out is a book that examines evidence that supports an alternative description of history from what has been taught by Western (Caucasian) teachings. History taught in many schools around the world has been distorted in order to support the false allegations that Caucasians are the only people who have made major contributions to history. For some reason, religion is the only subject that many people are willing to accept without proof, but that is a discussion for another time.

The underlying purpose of this book is to spark the reader's interest in learning the true history of the world. Hopefully, everyone who reads *Check This Out* will want to know more about what has been discussed in *Check This Out*. The information presented in this book is not even the tip of the iceberg when it comes to the mountain of information that the common person is kept from knowing or never told about.

The information in this book has been researched and collected by me over more than twenty years. Everything learned was gathered from the Internet, books, conferences, scholars, and personal experience. Originally, the information learned and gathered was used solely for my personal improvement. Yet after meeting many people who have no idea that they have been fed false knowledge, I realized a greater purpose to what I had been learning. Using my knowledge to help begin opening people's eyes to the truth became the focus of my learning.

Ancient history is important because it shows us what we have already accomplished and what obstacles and difficulties to avoid. Without Knowledge of our past, it is extremely difficult to choose a path that will take you forward in the future. It would be entirely possible to end up on the path that is headed back the direction from which we came. Knowing from where we came and what we accomplished in the past can help keep us headed in the right direction. If anything, this short glimpse at history should shine a bit of light on the direction from which we originated. The evidence I have gleaned from numerous sources points to the stars as the origin of intelligent life on Earth.

The subject of intelligent life on other worlds has been a subject of discussion for as long as people have gazed at the stars above. Some people not only believe intelligent life exists elsewhere in the universe, they believe that beings from these other civilizations have visited in the past and still may be visiting Earth. There are also a large number of people who do not believe that intelligent life exists elsewhere in the universe, at least not advanced enough to reach Earth (not more advanced than humans). To believe that Earth is the only planet where life exists is both naive and selfish. The arguments for both sides will be reviewed.

One justification by skeptics, used to support that we are alone in the universe, is that if there were aliens, why haven't we seen them and why don't they show themselves to everyone? These questions assume several things. First, one would have to believe that extraterrestrials look different from Earth people. Second, one would have to think that extraterrestrials would want to interact with humans. A third assumption is that since the news has not said anything about it, then it could not have happened. None of these assumptions are well thought out, nor are they substantiated by evidence. First of all, nobody has been everywhere on Earth and met every person on the planet. There are many animals, plants, and places on this planet that have never been seen by any person at all. No one should ever believe that just because they have not experienced something personally, it is impossible. In most cases, people do not believe anything, unless it is in the news. The news and religion are the only sources that the general public will believe without having to see any type of proof.

There are some people who question the existence of beings from other worlds because they have not seen these other-world beings. The argument that if you have not seen beings from another planet they must not exist is a naive perspective. A highly advanced society would most likely decide not to interact with a prejudiced, violent, less civilized society. It has been posed that humans on Earth are like small organisms in a petri dish being watched and studied by more advanced beings. Given the fact that we are basically unable to leave our home and explore the universe, we cannot begin to accurately speculate what lies in the infinite expanses of space. Speaking from a scientific point of view, I think it is a virtual certainty that life exists beyond this planet. In fact, it is ludicrous to think otherwise. There are more galaxies than we will ever know. Astronomers estimate that there are roughly fifty thousand stars in the Milky Way galaxy alone, which possess planetary systems that resemble the solar system in which we live. These neighboring planetary systems have properties that suggest that life (as we know it) could indeed be supported. Again, that is fifty thousand stars, and each of those could have multiple life-bearing planets surrounding them. The Kepler satellite has already identified over five hundred Earth-like planets. Also, keep in mind that these numbers are just looking at the

Milky Way galaxy. It is mathematically impossible for Earth to be the only planet with intellectual life-forms.

There are many scientists who look at the Drake equation as mathematical evidence that intelligent life exists elsewhere in our own galaxy. In 1961, an astronomer named Frank Drake came up with an equation to estimate the number of planets in the galaxy with life-forms that are advanced enough to communicate with Earth. Although this is one of the best equations in use to estimate the number of intelligent civilizations, it does not account for how long signals take to reach from one planet to another. By the time Earth receives a signal from another world, that other world could have completely changed. The other problem with the Drake equation is that it assumes other intelligent civilizations communicate with each other the same way humans do. Dolphins are more intelligent than humans, yet they do not use "technology" in the ways humans do. Dolphins use a higher percentage of their brain's potential than humans do. In the movie *Lucy*, an interesting question was posed, "Are humans more concerned with having than they are with being?" It would seem logical that a highly advanced intellect, whose concern is mental advancement, would not have much to discuss with a human.

Another issue that many people do not realize is the time discrepancy of universal images. Stars and other planets are extremely far from the Earth. The closest star to Earth is about thirty-eight trillion kilometers (just over four light-years) away, and that is the closest! Since the distance of galactic bodies was only going to increase, scientists decided to mainly use light-years in place of kilometers. One light-year is the distance it takes light to travel in one year, which is approximately 9,460,800,000,000 kilometers. Using light-years instead of kilometers, scientists can say a distance of 9,460,800,000,000 kilometers equals to one light-year. The issue is, if a planet in a solar system is five hundred thousand light-years away, the images received of that planet are from five hundred years ago. By the time the light gets to Earth and humans have the opportunity to collect it, an entire civilization could have developed on that distant planet.

Everything that is used today to detect life in the universe is designed to detect life like our own. With the immense variety of life on this planet, better yet, the immense variety seen in one species, why should every planet with life on it be the same? Should life be unable to take all forms? The form that life takes should only be limited by the number of possibilities, which is infinite. The viewpoint that all life must conform to the styles we have seen on this unimaginably tiny speck in the vastness of space is a horribly uneducated assumption. People on Earth, no matter how accomplished or intelligent, do not have the slightest idea what is going on elsewhere in the universe. The most brilliant minds in science still don't really have a clue about what is going on with civilization right here on Earth.

A renowned physicist, Dr. Michio Kaku, says that there are three classes of civilizations. The first type of civilization is a planetary civilization. This type, type 1, has the ability to control their entire planet. They control the rivers, the oceans, the volcanoes, the rain—basically all terrestrial events. Type 1 can either move their whole planet to another location in space, or they can move their whole civilization to another planet. The second type of civilization is called a stellar civilization. In addition to having mastered their planet, a stellar civilization has the power to control the output of a star itself. A type 2 civilization can reignite its star, if it faces cooling, and can even prevent the collapse of a star. A stellar civilization has the capability to harness energy from the stars themselves. The third type of civilization is considered to be a galactic civilization. A type 3, or galactic, civilization has the capacity to utilize the Planck energy. The Planck energy is the energy of nebula and black holes. This capability could possibly allow a type 3 civilization to control time. Although not as commonly accepted as the previous three types, there is a fourth type that has the ability to control dark matter or energy. Dark matter is thought to be what the universe is made of. It is the creative material for everything in existence. Overall, there are four types of civilizations that are theorized to exist among the cosmos.

Looking at the different types of civilizations, one must ask, which of the types is the human civilization? Do we control the fabric of space? Do we control the energy output of our sun? Do we control the events that occur here on Earth? The human civilization does not control the fabric of space, the output of a star, or the events that occur on Earth. Therefore, the human race is a type 0 civilization. Being a type 0 civilization means that we still have work to do in order to become the lowest-level civilization. Scientists, such as Dr. Kaku, think that we are actually relatively close to becoming a type 1 civilization. They estimate that within the next one hundred years, humans will be able to govern the natural events of Earth.

Every day, scientists work on finding new ways to predict what is happening right here on Earth. Collaborations from scientists the world over still fail to predict natural disasters. The events and aftermath of hurricane Katrina have shown just how powerless we are to predict events on Earth. There are still many places around the globe that are still unknown to modern society. Humans know even less about what is happening on the ocean floor. Ninety percent of the ocean is still unexplored. The ocean covers more of the Earth's surface than land, and yet, people believe they have a good understanding of the planet's happenings.

The deepest hole drilled into the Earth, by a human, was around 14½ kilometers down. Fourteen and a half kilometers below the surface is roughly nine miles into the Earth. Although this depth seems impressive, the continental crust of the planet is estimated to be between thirty and fifty kilometers (twenty and thirty miles) thick. The oceanic crust has been projected to be, on average, between five and ten kilometers (three and six miles) deep. The Earth's crust only accounts for 1 percent of the Earth's total mass. There are different scanners and devices that scientists use to determine what is beneath the crust. Since nobody has been to the center of the planet and come back to tell the public, we can't truly know what is down there. So what is under our feet is still technically a mystery. Humankind knows so little about the planet on which we live, yet those in charge refuse to accept and convey how little we truly know about the universe.

The Universe

   The universe—what more needs to be said? *Universe* is possibly the most enormous, underrated, and least understood word used in the human language. It is a word that is not given the respect that it deserves. The word *universe* should be on the same level as the word *god*. People often forget, or are unaware that the word *universe* represents everything in existence. There are scientists who have said that if someone was to count every grain of sand on every beach in the world, that number would not even come close to the number of galaxies (groups of solar systems) there are in space. Knowing that nobody can count the number of galaxies seen in the night sky means that there are more galaxies out there than we will ever know. If there are more galaxies than can be counted, then there are far more planets than we could ever imagine.

Astronomers estimate that there are roughly fifty thousand stars in the Milky Way galaxy alone, which possess planetary systems that resemble the solar system in which we live. These neighboring planetary systems have properties that suggest that life (as we know it) could indeed be supported. Again, that is fifty thousand stars, and each of those could have multiple life-bearing planets surrounding them. The Kepler satellite has already identified over five hundred Earth-like planets. Keep in mind that these numbers are just looking at the Milky Way galaxy. Scientists are beginning to admit that not only that life could very well exist in deep space but quite possibly in this very solar system as well. There are several celestial bodies in this solar system that have the basic elements to support the type of life with which we are familiar. Venus, Mars, Jupiter's moon Europa, and a few others are seriously being examined for signs of life.

There have been events that have caused debates among the scientific community. One such event is the WOW signal, which was detected by Ohio State University in 1977. The astronomer who received the thirty-seven-second-long signal wrote WOW on the data sheet printed from the telescope, which stuck as the name of the event. The signal originated from a location near the Sagittarius constellation. Radiation signals received from space normally cover a wide range of frequencies. The WOW signal was neatly within the range of radio frequencies banned for international use.

Venus

   Venus is the name given to the second planet from the sun. The Romans named the planet after their goddess of beauty and love. In ancient times, this planet was also called by some the morning star. Venus is the third brightest body in our known solar system. On clear nights, this planet is actually bright enough to cast shadows. Venus has an extremely thick atmosphere. The thickness of the atmosphere prevents us from being able to observe the surface of the planet. The fact that Venus's atmosphere blocks our ability to see the surface of the planet leaves the actual happenings on the surface of the planet to speculation.

   Venus is the closest planet to the Earth in the solar system. Venus is also closest to the Earth in its composition, size, and distance from the sun. It is said that the surface of the planet is covered with volcanoes that are constantly active. The thick atmosphere and continual volcanic activity have created a superheated environment. It is quite possible that some type of life-form has evolved on Venus. There are organisms on this planet that can survive extreme environments like boiling water, acid, ice, and even in dry rock, called extremophiles. Although no organisms have been found living in lava, there have been many organisms found living around volcanic vents.

Mars

Mars, our red celestial neighbor and the fourth planet from the sun, has been one of the most interesting lights in the night sky for as far back as recorded history. Mars, the light in the sky, has intrigued onlookers because of its reddish-orange glow, which sets it apart from its visible companions. The Roman Empire saw the reddish-orange glow in the sky and personified Mars as their god of war. Greeks associated Mars with fire and connected the planet to their god Aries. The current zodiac sign for March is Aries and the planet Mars.

There are ancient stories that tell of a city, called Cydonia, on the red planet from a distant past. Who are we to say that these ancient stories are not true accounts, as they claimed they were. Stories of a city on the red planet came from the people from Sumer—a civilization that had writing, schools, government, and religious practices thousands of years before KMT (Egypt). Sumer is the oldest nation on record to have, what is now considered, an advanced civilization. Sumer will be briefly discussed later in *Check This Out.*

It is speculated that Mars was once a planet much like Earth. Mars shows signs of having had oceans and rivers and great lakes. Signs of water, to humans, usually means that the planet most likely supported complex life. Signs of ancient oceans and rivers means that there could still be water, trapped deep beneath the surface of the planet. Satellite pictures of Mars show north and south poles that are capped in what looks like snow. A layer of frozen water rests beneath the permafrost on Mars's poles. The government tells us there is no liquid water on Mars, but that says nothing about underground. In 1976, radioactive methane gas was detected in the atmosphere of Mars. Radioactive methane gas is a byproduct of living organisms. NASA's *Viking* landers did experiments on the Martian soil and looked for the production of methane gas, which it found. Some scientists think that the production of the methane gas was from chemical processes and not biological ones.

If extremophiles from Earth could conceivably live in conditions such as we know exist on Mars, we certainly can't rule out the possibility that there's life on Mars today or that there was life on Mars when it was much warmer and wetter than it is at the present time.

—Richard Hoover, NASA scientist

Europa

Out of the planets and moons in the solar system we know, there is a handful that could potentially support life as we know it. Europa, one of Jupiter's moons, has been identified as one of the most likely places in our solar system for the existence of complex life. The surface of the planet is allegedly covered in ice. It has been reported by the National Aeronautics and Space Administration (NASA) that a vast liquid ocean lies beneath the frozen surface of Europa. Beneath the frozen surface of Europa, there is said to be twice the amount of water as there is on the entire Earth. That is a lot of water! If water is needed for life, then it should be a virtual certainty that life exists on Europa.

Scientists who are hopeful of finding life in our solar system are most certainly looking at Europa with great enthusiasm and optimism. The trip from Earth to Europa takes about thirteen months to complete. A thirteen-month trip, one way, is a short trip in celestial terms. Europa is the only place in our solar system that is known to have liquid water. The next closest star to us is far out of our reach. Humans do not even possess the technology to leave the solar system, much less reach another solar system. The presence of liquid water and distance from Earth make Europa the best option for exploration. Of all the planets in our solar system, Europa is the best hope for finding extraterrestrial life. Europa is also the best option for exoplanet human colonization.

The Klerksdorp Spheres

The Klerksdorp spheres are possibly the most controversial artifacts ever found. The Klerksdorp spheres are reported to be the oldest artifacts on Earth; these spheres have been dated at around 2.8 billion years old. These items were found in Klerksdorp, near Ottosdal, South Africa. During mining projects, workers were coming across hundreds of small spheres imbedded in the rock. A range of discs were found, from spherical to disc-shaped. The spheres measured roughly between 0.5 centimeters (cm) and ten centimeters in diameter. Many of them have one or several grooves running around their equators. If these spheres were naturally made, it is the only instance, in the known world, where spherical balls, with naturally occurring, evenly spaced rings around the center, appear imbedded in a different type of stone. Another explanation has been proposed. The spheres could have been made by intelligent hands. The dimensions and the evenness of the grooves are too precise to have been made without help from an intellect. It was reported that the spheres were found embedded in pyrophyllite deposits. The spheres themselves come in two type;, the first is a hollow ball filled with an unknown spongy, fibrous white material, and the other is composed of a bluish metal with white flakes.

P i r i   R e i s   M a p

The Piri Reis map is one of the most interesting artifacts in history. The Piri Reis map was the possession of Admiral Piri Reis of the Turkish military. The admiral found the map while in China and stated that the map, drawn in AD 1513, is a recreation that comes from an even older version. The map shows the East Coast of South America, the West Coast of Africa, and the North Coast of Antarctica. At the time that the map was created, the coastline of Antarctica was not visible. It has been estimated by scientists that the last time the Antarctic was free of ice was between twelve thousand and seventeen thousand years ago. According to western teachings, two men of the British navy, William Smith and James Bransfield, discovered and began mapping Antarctica in 1820. Who had the technology to map out Antarctica before it was covered by ice?

Machu Picchu

    An ancient hidden megalithic city lies in the jungles of South America. It is called Machu Picchu. Machu Picchu, "old peak" in Quechua, is an amazing city built high in the mountains of Peru. Entirely invisible from the ground, this city was built just over nine thousand feet above sea level on the top of a mountain. The city is roughly five square miles and is completely self-sufficient. There is evidence that this mountaintop city was used by the Inca. Although the city shows signs of use by the Inca, the natives say that the city existed long before the Inca arrived.

Machu Picchu exhibits the same engineering skill as many of the other ancient megalithic sites around the world. Some of the stones used in the structures weigh over fifty tons, yet they were put together with the utmost precision. The builders were able to create houses and terraces without the use of mortar. The entire city was built so that nothing was needed from elsewhere. Terraces were built for growing food, while natural springs kept the city watered. Whoever built Machu Picchu perfectly incorporated the environment into the design for construction. How was ancient man able to construct such magnificent cities without the use of advanced technology?

Tiahuanaco (Tiwanaku) and Puma Punku

Another one of the most spectacular and bewildering megalithic sites in the world is the city of Tiahuanaco. Tiahuanaco is considered the Baalbek of the Americas and is possibly the most enigmatic site in the world. It is positioned almost exactly between lakes Titicaca and Aullagas. Tiahuanaco sits at an elevation of around thirteen thousand feet above sea level. Many people consider this area to be the Tibet of the Americas due to the high elevation and barren, arid environment.

Who build the city and when are just two of the questions plaguing archaeologists. The age of Tiahuanaco has puzzled researchers due to it being estimated at over seventeen thousand years old. Legends say that the temple used to lie on the shore of Lake Titicaca. Scientists agree that the waters of Lake Titicaca recede about 0.125 inches per year. The shore of the lake is about eighteen miles from the temple today. If the temple was on the bank of the lake, it was built around one hundred forty thousand years ago, considering how far the lake is from the temple and the rate at which the lake recedes. A construction date of 140,000 BCE would make Tiahuanaco the oldest structure in the known world and completely change history as it is taught.

Beyond the age, there is the extremely sophisticated design and construction. Some of the stones used in construction weight up to two hundred tons. The blocks, composed of andesite, were cut with what appears to be machine precision. The closest source for andesite is around twenty kilometers to the north. No mortar was used, and yet knife blade cannot penetrate the cracks. There were also reports of giant chairs being found that weighed around ten tons. Architecture of this sophistication was not supposed to exist at that time, according to Western teachings.

There are many theories regarding how and why this city was built. The socially accepted idea is that the city was built by a massive human workforce. Indigenous people of the area say that it was the first city in the world and that it was built by a race of giants. Artifacts found at Tiahuanaco do support the story of the indigenous people.

Nazca Lines

South America is riddled with amazing archeological sites. Peru is one of the most famous countries in South America for incredible archeological locations. When viewed from the ground, one sees thousands of stones that appear to be curves and extremely straight lines running off into the distance without any apparent purpose. The lines of stones are known as the Nazca Lines. In order to fully appreciate what was done by the creators of the lines, one must see them from the air. It appears that hundreds of thousands of stones have been strategically placed on the ground in patterns to make giant pictures only recognizable from far above the land. It is obvious that these pictures were meant to be seen by someone looking down from a great height above the ground. The Nazca Lines located in the Peruvian landscape have baffled both scholars and tourists.

Jupiter Temple and the Stone of the South

The Jupiter Temple in Baalbek, Lebanon, is arguably the most amazing structure built by intelligent hands. The Roman Empire claimed responsibility for building the temple, although the evidence tells a different story. The Romans did not have the technological capability to cut and move the enormous blocks that were used in the foundation of the temple. When you look at the outer walls of the Jupiter Temple, you will notice the two distinct levels of weathering on the building. The structure shows signs of two different building times. It appears as if the Romans came across what was left of an ancient structure and added what they needed in order to utilize the building. The Jupiter Temple structure contains the largest stones cut and moved by any people in history.

There are three stones used in the construction known as the *trilithon stones*. These three stones weigh around nine hundred tons each. As if that were not impressive enough, there are roughly twenty-four stones that weigh about three hundred tons each. Two additional stones were discovered near the building site that were cut but not used. One of these stones is estimated to weigh around one thousand tons while the other has been estimated to weigh around one thousand two hundred tons.

The Stone of the South is the largest stone cut and moved on Earth. The Stone of the South weighs an amazing two thousand tons.

M o a i

There is a tiny island off the coast of South America located in the Pacific Ocean. The island is called Rapa Nui by the indigenous people but is known to most people by its European name, Easter Island. Rapa Nui is under the control of the Chilean government and is located roughly three thousand six hundred kilometers or two thousand two hundred miles off the mainland. With only 171 square kilometers or sixty-six square miles of land, Rapa Nui is basically a speck in the Pacific Ocean. Rapa Nui could be the most remote inhabited island on Earth.

Easter Island, or Rapa Nui, is also known by other lesser known names like Te-Pito-O-Te-Henua, which means the Navel of the World, and known as Mata-Ki-Te-Rani, which means Eyes Looking at Heaven. The ancient names and the host of mythological details ignored by mainstream archaeologists point to the possibility that the remote island may once have been both a geodetic marker and the site of an astronomical observatory of a long-forgotten civilization. The island radiates a strange electromagnetic field that interferes with electronics. Although the electromagnetics on the island are mystifying, the island is famous for the giant stone sculptures, called Moai, that have been erected.

There are 887 Moai positioned around the island. The largest Moai, called Paro, stands almost ten meters (thirty-three feet) tall and weighs an impressive eighty-two tons. The heaviest Moai on the island, a shorter squatting Moai at Ahu Tongariki, weighs eighty-six tons. There is an unfinished Moai still in the quarry that would have been roughly twenty-one meters (sixty-nine feet) tall, with a weight of about 270 tons. All the statues seem to have been cut from the same quarry and transported to their individual locations on the island. A total of 834 of the statues are cut from tuff, compressed volcanic ash, from the Rano Raraku volcano. The smallest number (thirteen) of the Moai are carved from basalt, twenty-two from trachyte, and seventeeb from fragile red scoria. It has been suggested by some scientists that these great statues were moved to their locations with the use of wooden rollers. Speaking from a scientific point of view, the roller theory is completely unsubstantiated. The shape of the Moai is too awkward for the use of wooden rollers, not to mention there are too few trees on the island. Another mystery is how the large hat-like stones were placed on the heads of the Moai. There have been no publicized theories as to how the stones were placed on the heads. These are a few of the mysteries that shroud Rapa Nui.

Pyramids at Giza, Egypt, Africa

Possibly the most well-known ancient structures in the world are the pyramids at Giza, KMT (known today as Egypt), Africa. Situated in a vast expanse of desert there lies one of the greatest engineering feats known to mankind. It has been accepted by some people that the pyramids were built around 2,500 BCE, each by a different pharaoh. One could ask though, did each successive pharaoh simply agree to follow the previous pharaoh's plan for his burial? Did each successive pharaoh simply agree to have a smaller pyramid built for them, or were they just running out of supplies? And more impressively, how did they, at that time with the technology they allegedly used, know exactly where the center of all the land mass on Earth was? A lucky coincidence? Another question is this, does it sound logical to have a bunch of unskilled people build anything of importance? Allowing new data and interpretations may help find a real answer to the mystery of the pyramids.

There is new astronomical data that suggests that the pyramids were more likely built around 10,500 BCE. The position of the pyramids perfectly line up with the constellation Orion right around the year 10,450 BCE. Technology of the pyramids should not have existed anywhere near 10,000 BCE, according to traditional American teachings. Scientists agree that Hermaket (the Sphinx, the Greek name) was built much earlier than the alleged 2,400 BCE date for the pyramids. It has been suggested that Hermaket was built around the 10,500 BCE date. Is it possible that the pyramids and Hermaket were built at the same time? People with the knowledge and ability to build some of the most advanced structures on Earth might have been able to build several structures simultaneously. Ancient KMT should never be underestimated when it comes to feats of great ingenuity.

Saqqara Bird

An object of great debate is the Saqqara Bird. The Saqqara Bird was found among the possessions of Ta-Ji-Imen (Ptah-Dji-Amen), when his tomb was vandalized and robbed in 1898 by European thieves. Ptah-Dji-Amen's grave was located in Saqqara, Egypt, North Africa, about ten miles southeast of Giza. The Saqqara Bird model recovered from the grave has stirred something of a controversy. To an untrained eye, the object is a mere toy or piece of art. The simplest description of the object is a small bird with outstretched wings. The obvious appearance has led many people to classify the object as either a ceremonial object or a child's toy. It has also been speculated by some that the Saqqara Bird served as a weather vane. Others have come to the conclusion that the object might have been a model for some type of flying device or vehicle.

According to history (Western Society), KMT (Egypt) did not have the know-how or the technology to develop flying machines or any complex machinery. The thought that ancient Egyptians were technologically inferior or inferior in any way, compared to modern-day society, is a complete misconception. The most compelling pieces of evidence that prove KMTic ingenuity are the Great Pyramids at Giza. How could anyone look at the pyramids in present-day Egypt, whose construction is still a mystery, and question the level of scientific and technological understanding of these Black people? The science of flight as a whole is less complex than the science needed to construct the complex at Giza.

Due to the theories that the Saqqara Bird could have been used as a model for a plane, several specialists decided to make replicas of the Saqqara Bird and test it for flight stability. Martin George, a model plane builder, tested a recreation of the model made out of Balsa wood. He said that his version of the model was unable to stay aloft. The original Saqqara Bird does not have the typical tail stabilizer attached to it. Even when the missing tailpiece was added to the model, George said that the flight was unstable and, therefore, would not be suitable for any type of actual flying vehicle. On the other hand, there are many who believe that the model is an actual miniature version of an actual flying device. A man named Dr. Dawoud Khalil Messiha, an Egyptian physician, was so intrigued by the model that he decided to test the flight capabilities of the model bird. He noticed that the Saqqara Bird artifact did not have the typical tail stabilizer seen on birds but rather a vertical tail stabilizer as seen on modern airplanes. Dr. Messiha reported that his version of the Saqqara Bird did indeed achieve stable flight. Dr. Messiha said that his model only flew when an additional tail stabilizer was implemented, which he believed was part of the original artifact and was lost at some point in time.

Ancient Model Planes

Americans have been taught that the Wright brothers were the first people to invent the airplane. The typical approach to teaching history is that Europeans were the creators of everything important, good and great. In America, and in most other countries, schools primarily teach children and adults about European culture. Europeans, contrary to what is usually written in history books, were not the first to devise flying technology.

These model planes were discovered in South America. There are stories that tell of how the ancient deities could travel through the air in special vehicles. Some of the flying vehicles from ancient South American stories tell of giant flying turtle shells. Could ancient man have been witness to airplanes and rocket ships and even flying saucers? There are numerous stories, pictures, and models that document the fact that people did fly—thousands of years before the Wright brothers.

In Africa, there is a building in Upper KMT (a.k.a. Egypt) known as the Temple of Abydos. The king of Abydos was known to the people as the Serpent King. In his temple, there are hieroglyphics of what appear to be flying vehicles. The hieroglyphs have been dated as being around five thousand years old.

The Olmec Culture

Many people have been taught that Christopher Columbus discovered America. Reporting that Columbus discovered America is a complete lie. First of all, how does someone discover a place that already has people living there? The truth is that Columbus never set foot on North America at all. Whether Columbus landed on North America or not is irrelevant. The important point is that there were already people who inhabited the continent. Who were these indigenous Americans, and where did they come from? We have been shown images of indigenous Americans pictured with Caucasian features. Society has been pushing the image of only the indigenous Americans who uphold the myth that Caucasians landed in the Americas before anyone else.

History as told by Europeans requires everyone believe that language, arts, and science were creations of European cultures—another utter and complete lie. Europeans learned everything that they knew, of any worth, from their interactions with African cultures. Pathagerus and Leonardo da Vinci, to name a few, admitted learning from the Egyptian teachings. Sometimes people forget that Egypt is in Africa.

Nestled in the jungles of Central America are some of the oldest and most amazing archeological sites in the world. Located on the southern Gulf Coast of Mexico, in the areas known today as Veracruz and Tabasco, one of the most ancient cities can be found. San Lorenzo Tenochtitlan, Mexico, is home to the first civilization in American history. The first American civilization is known today as that of the Olmecs. The Olmecs were estimated to have been a flourishing culture around 4,000 BCE. Central America has the oldest faces that are carved out of stone, documenting who are the original indigenous people of the Americas. Scattered through the jungle are giant stone heads. The faces on the statues are obviously of African origin. The broad, flat noses and full thick lips are clearly recognizable. Interestingly enough, black was considered to be the color of the elite, a symbol of high society.

According to the most recent archeological information, the Olmecs were America's first settlers and builders of its first advanced civilization. Olmec technology was advanced, in some ways even more than ours. Their cities included paved roads, complex drainage systems, irrigation systems, and buildings that still stand thousands of years after they were built. The Olmecs developed and introduced the first calendar in the Americas. It has been determined that the Olmec calendar was in use in the year 3114 BCE. The giant stone heads, scattered through the jungle, were each cut from a single piece of basalt stone. The basalt used for the sculptures is found in the Tuxtlas Mountains, roughly eighty miles away. These statues were cut and then carried for about eighty miles through a dense, swampy jungle to their resting places. Some of the stone heads weigh up to fifty tons. The Olmec heads have caused a great controversy due to the level of engineering knowledge needed to accomplish such a great feat.

Evidence has been presented that links the Olmecs of Central America to the Dogon of Mali, Africa. The physical features of the Olmecs are clearly African in origin. The language and religious practices of the Olmecs is that of the Mende or Shi people from West Africa. Some sources, the Gladwin Thesis, have estimated that Black Africans embarked on a maritime voyage around the world sometime near 73,000 BCE.

Tombstone of Pical

In present-day Chiapas, Mexico, there is a building that houses an ancient mystery. The building containing the mystery is known to the public as the Temple of Inscriptions. Buried deep inside the Temple of Inscriptions is a tomb that defies modern explanations. The tomb is said to be the resting place of the Mayan ruler K'inich Janaab' Pakal, ajaw of Palenque, now known as Pical. The legend of the deity says that Pical came to Earth from a distant planet. It is said that he came to Earth and taught the people here advanced agriculture and building skills. The most amazing part of this story is the cover of Pical's tomb.

On the cover of Pical's tomb, there is a carving of what appears to be a man sitting in some sort of vehicle. The vehicle resembles what we might call a rocket ship or some other kind of flying machine. The man, in the carving, appears to be sitting on a highly cushioned seat. He also appears to be wearing what looks like a tightly fitting uniform with possibly a utility belt. Pical is intently staring ahead as if he were driving or flying a vehicle. He appears to be working some gears or switches with his hands and maybe using a pedal with his feet. There are visible flames coming from the back of his vehicle.

The carving on Pical's tomb is quite intriguing, to say the least. What was it that the artist saw that inspired the carving? Could the artist have seen Pical come or go in a vehicle that traveled into the heavens? Pical's tomb was carved long before any flying vehicles were thought to exist (by Western culture). It would be foolish to completely dismiss an ancient story as fiction, when there is an artifact that exists to backs up the story. It seems as though the ancient story could very well be, in fact, ancient history.

Utah Petroglyphs

Some of the most interesting ancient cave paintings found anywhere on Earth are found in the United States of America, in Utah. As with most cave paintings, drawings were made using things that the artist had seen in his/her environment. Strangely, there are many pictures that do not conform to anything that the artist could have personally seen in the surrounding area. There are disc-shaped objects and many abnormally shaped beings. Keeping to the idea that ancient man drew what he saw, many people have been left to speculate as to what or who the figures in these drawings represent. Some people speculate that the strange figures represent gods or spirits. If the petroglyphs do, in fact, represent gods or spirits, the question is, why did early man decide to depict these beings in a form other than that of a human? To this day, many of the paintings are still a mystery.

The Great God Mars

In Southern Algeria, Africa, there is an area where there are some of the oldest cave paintings in the world. The caves and cliffs of the area known as Tassili n-Ajjer are riddled with paintings. Scientists have dated the paintings as drawn before 20,000 BCE, as early as humans have inhabited the area. Ancient cave painters have consistently depicted what was seen in the environment around them. Ancient cave art has been used as a guide to determine what was going on in the area during the time that the painter was alive. Animal and vegetable life was drawn with great detail to proportions and shapes. In short, ancient people drew what they saw the way they saw it.

With the knowledge that ancient people drew what they saw, there are some paintings in the area that defy explanation. First of all, there are paintings of elephants, gazelles, trees, and flowers. These pictures are found in one of the most inhospitable places in the world, an expansive desert. Beyond the out-of-place animals and plants, there are pictures of what appear to be modern objects in design. Some of the pictures resemble zippers, backpacks, helmets, antennae, and strange beings. There are drawings of beings with abnormally large heads and bodies. Some researchers say that these abnormalities are simply representations of ceremonial masks and clothing. Other researchers think that these figures represent a different type of being that ancient people encountered.

Uzbekistan Cave Painting

Located just northwest of China, in the country of Uzbekistan, there exists one of the world's greatest cave paintings documenting the visitation to Earth of beings from another planet. The Uzbekistan cave paintings were found in a cave in the Fergana Valley and depict a being in the foreground, partially suited holding a disc, a being, in the background, fully suited, and what looks like a flying machine with mountains in the background. The age of the painting has been estimated by some as having been created around 8,000 BCE.

The main figure is dressed in what appears to be some type of suit. A domed helmet can be seen partially on the being's head. The being has the appearance of what could be interpreted as a reptilian face with a feathered head or a demonic face with sharp teeth and possibly flames coming off its head. The left hand is wearing a thick insulated glove and is holding some sort of disc with a spiral pattern on it. The right hand is uncovered and is making the gesture, in this day and age, recognized as the OK symbol. The suit has hoses and gadgets on it. The suit's design suggests travel into extreme environments. On the back of the being, there is a wing that resembles the wings of a bird or an angel.

The second being, in the picture, is fully suited. It is the basic, and classic, image of a "spaceman." The figure wears a helmet that has large eyes and antennae sticking up out of it. Above the being, in the background, is what looks like a disc with possibly smoke or gas coming from the bottom. There are also three circles in what could be considered the sky. What could these circles represent—the sun, the moon, and what else? There is no agreement on the explanation of the images on the Uzbekistan cave painting.

Dropa Discs

In 1938, an archeologist named Chi Pu Tei uncovered rows of little graves in the Baian Kara Ula Mountains near the Sino-Tibetan boarder. One report said that bodies uncovered ranged in height from two feet to four feet tall. The graves were in a cave positioned under drawings of what appear to be stars, planets, and beings with helmets. Among the graves, archeologists have discovered over seven hundred disc-shaped objects buried along with the bodies. Some of the discs were analyzed to determine their composition. The discs were mainly composed of granite. Surprisingly, the discs contained large amounts of cobalt and other metallic elements. When placed on an oscillograph, a hum, or vibration emanated from the discs as though they carry an electric charge.

The discs have a small spiral groove running from the outer edge to the center, like a modern-day vinyl record. Upon closer study of the disc, it was discovered that the grooves were actually tiny glyphs. The glyphs were studied and translated by Professor Tsum Um Nui of the Academy of Prehistoric Research in Beijing, China. The translation told of a group of scientists from another planet who were investigating Earth when their vehicle crashed into a mountain. Unable to repair their vehicle, the beings attempted to comingle with the indigenous inhabitants of the area. Initially, the indigenous people did not accept the attempt by these beings to comingle. They were afraid of the odd-looking beings and killed many of them. After a period of time, the beings were finally accepted. The beings eventually interbred with the indigenous inhabitants.

The story translated from the discs is an amazing one. The story is more remarkable considering the fact that the discs have been dated at 12,000 BCE. That's around fourteen thousand years old! All the evidence that was found in the area supports the story. The main piece of evidence people want to see is the spacecraft. If the craft was found, the government has not shared the news (which is completely expected).

Sumerian Culture

The oldest accepted civilization in known, or told, history is that of the Sumerians. Sumer has been dated at being completely developed by the year 4,000 BCE. Today's society is largely based on the structure of the Sumerian society. The Sumerians have been credited with developing the first form of written language. The Sumerians called themselves "Black-headed people" in their texts. In Sumer, great care was taken to record historic, scientific, governmental, and religious events. Researchers are finding it difficult to find evidence that explains the sudden appearance of advanced technology in Sumer.

One major question plaguing archeologists is how did hunter/gatherers spontaneously jump to building complex structures and an advanced government? Archeologists have found a multitude of clay tablets and stele (an ancient stone slab or pillar usually with inscriptions) with amazing historical recollections. According to what has been deciphered from these tablets and stele, the people of Sumer had help in reaching the sophisticated level of civilization they had achieved. Records tell of how advanced beings came to Earth from a distant planet called Nibiru. The beings from Nibiru, currently referred to as gods, came to Earth to collect supplies and minerals to repair the depleting atmosphere on their home planet. Due to the fact that these beings were so much more advanced, and they claim responsibility for creating the human race, they became known as deities. Sumerian deities played an active role in everyday life. There are texts that describe how humans were created using genetic engineering. According to these texts, a master scientist from Nibiru named Enki fashioned humans by combining his own DNA with that of an ape-man (possibly Neanderthal man). If scientists could simply accept what was written by the Sumerians, many of the questions regarding human evolution and other unexplained happenings would be answered.

Closing Remarks

In closing, I will share my conclusions regarding the information in *Check This Out*. My theory is that the universe is teaming with advanced civilizations. At some point, in the distant past, highly advanced beings from another planet came to this planet and found the conditions favorable to their purposes. I think that many of the megalithic structures around the world were either built by extraterrestrials or by humans with the aid of extraterrestrials. In most cases, the simplest explanation is the correct one.

Some of these beings who came to Earth were black in complexion, not brown but deep-space black. While on this planet, the beings found a creature, something between an ape and a human, who was similar enough to the beings who came to Earth so that genetic blending was possible. The first humans were the result of the blending. The reason "black" genes are the most dominant is because they are the original blueprints. The beings then taught humankind all that it knows. Humans idolized the beings for the knowledge and power they possessed. These beings became the first gods and goddesses in human history. At some point, the beings decided to retract their physical presence from the Earth. Some of the beings stayed and interbred with the humans. I believe that many different types of beings now come and go from the Earth.

The issue I have with the theory that humans evolved from apes and built the megalithic structures around the world is that there should be structures showing a gradual increase in complexity. If, as this theory suggests, humans built the megaliths, it would have taken generations to complete a project. My problem with the theory that everything was created instantly by some supernatural being is that evolution is proven and is always occurring. Many people do not believe in evolution because they cannot see it taking place. Evolution is a very slow process that takes place over many generations. When a plant, animal, or human adapts to survive in an environment, it is an example of evolution. The human life span is usually not long enough to witness any major changes in the world, much less the universe.

As far as it goes with most people's knowledge (even those with the highest degrees) about what is going on in the universe, I would have to say we know practically nothing.

Humans are still perplexed by what is happening right here on the planet we live on. How can anyone who has not personally been to a place call themselves an expert on that place? If you have not been to Mars, you cannot say, definitively, that there is nothing living there. Although the government may say there is nothing living there, ask yourself this: is the government always 100 percent honest? The government has revealed that a solar system, Gliese 581 system, does have an Earth-like planet. The Gliese 581 system is only 20.3 light-years (192 trillion kilometers or 119 trillion miles) from Earth, which is relatively close. The government has not disclosed whether or not life has been detected in the Gliese 581 system, and I doubt they will.

As far as life on Earth goes, if a place has water, then that place has life. If all the places on Earth that have water have life, why would anyone think it is a different case on other planets? Water has been discovered in various places in this solar system. It has been reported by NASA that a large amount of water ice is present on the moon. If the moon has water, is it too much to say that there could be life on it as well? Several of the moons of Jupiter and Saturn have also been determined to have vast liquid oceans. Life cannot be limited to Earth—some of the proof is in this book.

Thank you again for reading *Check This Out.* I truly hope that this book has been informative, as well as interesting. Above all else, I hope that this book has sparked your interest in ancient and alternative history. Never trust any source as a stand-alone authority. Research everything you have been told and taught. Nothing contained in this book should be simply accepted; everything should be <u>checked out</u>! With a better understanding of where we come from, we as a people can see what direction we should be headed. If there are any mistakes or if any of the information in this book is incorrect, please e-mail corrections along with questions and comments to etheoryal@gmail.com. Due to copyright issues some pictures could not be included in my book. They are available for viewing in my presentations. Contact me at etheroryal@gmail.com to schedule a presentation on Check This Out and see all of the pictures.

Peace and blessings to all.

Amon Gooden, BS

Sources

1: **Articles, Reports, and Studies in Various Issues of the Following Magazines, Newspapers and Periodicals**

Discover, 2012

Indianapolis News, November, 1975

Nature, December 17, 1891

2: **Individual Studies, Works, and People**

Dr. Delbert Blair

Mr. Eugene Braxton—personal acquaintance

Daniken, E. (1973). *Gold of the gods*

Dr. Maria Gooden—personal acquaintance

Dr. Warren Gooden—personal acquaintance

Holy Bible

Mr. David Ike

Dr. Michio Kaku

Mahabharata

Mr. Credo Mutwa

Ramayana

Sabloff, J. (1989). *The Cities of Ancient Mexico: Reconstructing a Lost World*, New York: Thames and Hudson Inc.

Dr. Saitsew, W. (1968). *The Dropa*

Simons, G. (1994). *Iraq: from Sumer to Saddam*, New York: St. Martin's Press, Inc.

Sitchin, Z. (1976). *12th Planet,* New York: Stein and Day

Sitchin, Z. (1980). *Stairway to Heaven*, Avon Books

Sitchin, Z. (1985). *Wars of Gods and Men*, Avon Books

Sumerian Deluge

Mr. Bill Witacker (Billel)—personal acquaintance

Dr. York, M. *Existence: How and Why*

Dr. Malachi Z. York—personal acquaintance

3: **Lectures, Videos, and Documentaries/Movies**

Arntz, W., Chasse, B., and Vicente, M. (Producers). (2004). *What the Bleep Do We Know!?* (Motion picture). United States: Roadside Attractions

Burns, K. (Producer). (2010). *Ancient Aliens* (Television broadcast). United States: Prometheus Entertainment

Cooper, W. (1972). *U.S. Navy Involvement and Briefing Agenda* (DVD). United States

Greer, S. (Director). (1993). *The Disclosure Project* (DVD). United States

Reinl, H. (Director). (1970). *Chariots of the Gods* (Motion picture). Germany: West German Films

Spielberg, S. (Directed). (2008). *Indiana Jones and the Kingdom of the Crystal Skull* (Motion picture). United States: Paramount Pictures

Walz, J. (Producer). (2008). *U.F.O. Hunters* (Television broadcast). United States: Motion Picture Production Inc.

4: **Places**

Iraq Museum International

Museo Nacional de Antropologia (National Museum of Anthropology, Mexico City, Mexico)

The University of Pennsylvania Museum of Archaeology and Anthropology

Printed in the United States
By Bookmasters